MathStart®

SUBTRACTING TWO-DIGIT NUMBERS

SHARK SWIMATHON

by Stuart J. Murphy

illustrated by Lynne Cravath

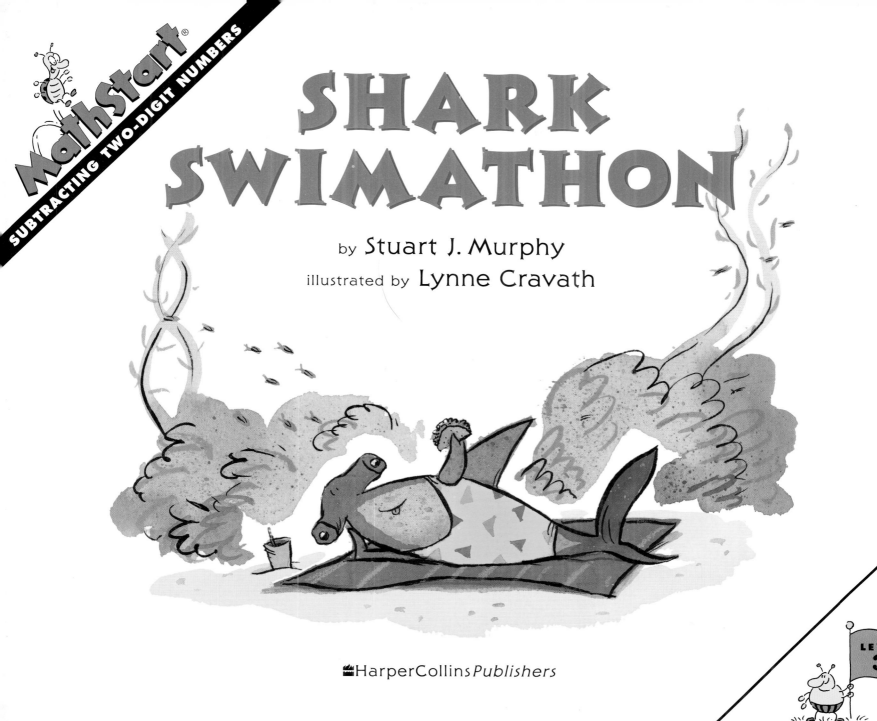

HarperCollins Publishers

LEVEL
3

To the Shedlosky Sharks—
Rene and Tom, Tara, John, and Eric
—S.J.M.

For Chloe and Jeff
—L.C.

The publisher and author would like to thank teachers Patricia Chase,
Phyllis Goldman, and Patrick Hopfensperger for their help in
making the math in MathStart just right for kids.

HarperCollins®, 🐟®, and MathStart® are registered trademarks of HarperCollins Publishers.
For more information about the MathStart series, write to
HarperCollins Children's Books, 1350 Avenue of the Americas, New York, NY 10019,
or visit our website at www.mathstartbooks.com.

Bugs incorporated in the MathStart series design were painted by Jon Buller.

Shark Swimathon
Text copyright © 2001 by Stuart J. Murphy
Illustrations copyright © 2001 by Lynne Cravath
Printed in the U.S.A. All rights reserved.

Library of Congress Cataloging-in-Publication Data
Murphy, Stuart J., 1942–
    The shark swimathon / by Stuart J. Murphy ; illustrated by Lynne Cravath.
        p.   cm. — (MathStart)
    "Level 3. Subtracting two-digit numbers."
    Summary: As members of a swim team do laps to qualify for swim camp, readers can practice
subtracting two-digit numbers to see how many laps are left to go.
    ISBN 0-06-028030-1. — ISBN 0-06-028031-X (lib.bdg.) — ISBN 0-06-446735-X (pbk.)
    I. Subtraction—Juvenile literature.   [I. Subtraction.]   I. Cravath, Lynne Woodcock.   II. Title.
III. Series.
QA115.M875   2001   99-30312
513.2'12—dc21   CIP
                                                                                                                              AC

1 2 3 4 5 6 7 8 9 10
❖
First Edition

After practice on Monday, all the Ocean City Sharks could talk about was state swim camp.

"We'd get to meet the best swimmers around," said Gill.

"And we'd learn a lot about swimming," added Stripes.

Gill and Stripes were the co-captains of the team.

"But we'd need money to get there," Tiny pointed out. "And we'd need money for lunches!" said Flip and Flap, the hammerhead twins. They were always hungry.

"We'd need a lot of money," Fin said. "And we just don't have it."

Just then Coach Blue swam in, waving a copy of *The Ocean City News.* "Here's our chance to get to swim camp!" she said. "To celebrate its seventy-fifth anniversary, the Ocean City Bank has made a special offer: It will send any swim team to camp if the team swims a total of 75 laps by the end of the week."

"That's a lot of laps," said Tiny doubtfully. "And we only have 4 days left."

"But there are 6 of us," Fin pointed out.

"I know you can do it if you all work together," Coach Blue told them.

Tiny took a deep breath and shouted, "We're Sharks, and we're fearless!"

"All right!" the rest of the team cheered.

On Tuesday, while the team warmed up, Coach Blue wrote their goal on a big sign.

SHARK SWIMATHON
Laps to swim:
75

Then she blew her whistle, and the Sharks started swimming. All 6 Sharks swam 1 lap, the length of the pool and back. Coach Blue shouted, "Go, Sharks, go!"

Tiny, Fin, Flip, and Flap all stopped after 2 laps. But Gill and Stripes kept on swimming. They stopped after 1 more lap. "Great swim!" Coach Blue shouted.

The Sharks watched as Coach Blue totaled their laps on her clipboard. Then she subtracted the total from the 75-lap goal.

"Super job!" Coach Blue announced.

"Now we just have 61 laps to go," said Flap as she and Flip gobbled down a snack.

Tuesday's Tally
Fin //
Tiny //
Flip //
Flap //
Stripes ///
Gill ///
total: (14)

13

On Wednesday the Sharks couldn't wait to get into the pool.

"Sharks to the water!" Coach Blue yelled as she blew her whistle.

The Sharks swam as hard as they could. Tiny stopped after 2 laps, but all the others swam 3.

Wednesday's Tally

Fin |||
Tiny ||
Flip |||
Flap |||
Stripes |||
Gill |||

total: 17

Then they all crowded around as Coach Blue again totaled their laps on her clipboard. When she finished, she subtracted Wednesday's total from the number of laps they still needed to meet their goal.

"We have 44 laps to go," announced Coach Blue.

All the Sharks arrived early on Thursday.

"I hope we can make it," said Stripes, looking at the sign.

"I hope so too," Gill said.

After the warm-up, Coach Blue shouted, "Let's go!" and the Sharks dived in. This time no one got out after 2 laps. Everyone swam 3, except Gill. This time he swam 4.

When they were done, the team waited while Coach Blue totaled the results. Then she subtracted the total from the remaining laps in their goal.

"Great going," Coach Blue said. "You did really well. We only have 25 more laps to go."

As the Sharks rested, Flip said, "I wonder what kind of food we'll get at swim camp. Maybe tacos."

"Tacos are my favorite," said Flap.

"Mine too," agreed Flip.

Thursday's Tally
Fin |||
Tiny |||
Flip ||||
Flap |||
Stripes ||||
Gill ||||
total: (19)

On Friday everybody warmed up and was ready to go. Everybody, that is, except Gill.

"Where could he be?" asked Fin.

"Here I am," groaned Gill as he swam slowly in.

"What happened?" Fin cried.

"I fell off my bike," Gill explained sadly. "The doctor said I can't race for a whole month!"

"Oh, no!" Tiny and Stripes moaned.
"There goes swim camp," sighed Flip.
"There go the tacos," sighed Flap.

The Sharks gathered around Coach Blue.

"We can't do it without Gill," said Tiny, taking off her goggles.

"Oh, yes, we can," said Stripes firmly. She looked at the sign. "We can make it if we each swim 5 laps."

"Remember," Gill said, "we're Sharks . . ."

". . . and we're fearless!" finished Tiny, putting her goggles back on.

Gill watched and cheered as the other
Sharks attacked the water.

Stripes finished her 5 laps in record time.

Flip and Flap finished their laps next.

Fin struggled to the end of his fifth lap.

Tiny was the only one left swimming.

"Go, Tiny!" the rest of the team chanted.

"Do it for Gill!" Stripes called.

"Do it for the tacos!" Flip and Flap yelled.

Finally, with a big gasp, Tiny made it to the end of her fifth lap. All the swimmers jumped back in the pool. "Way to go, Sharks!" Gill yelled.

27

Coach Blue blew her whistle to get the team's attention. "Don't you want to see the results?" she asked.

The Sharks all rushed over to the sign. Coach Blue had already put up the team's numbers.

"Swim camp, here we come!" Stripes called.

"Tacos, here we come!" Flip and Flap cheered together.

Friday's Tally
Fin ⊬⊬⊬
Tiny ⊬⊬⊬
Flip ⊬⊬⊬
Flap ⊬⊬⊬
Stripes ⊬⊬⊬
Gill
total: 25

The next day the Sharks made the sports page of *The Ocean City News.*

And the next month, the whole team made it to swim camp.

In *Shark Swimathon*, the math concept is subtracting two-digit numbers. Learning to subtract numbers that have more than one place value (ones and tens) prepares children for subtracting larger numbers.

If you would like to have more fun with the math concepts presented in *Shark Swimathon*, here are a few suggestions:

• Read the story together and ask the child to describe what is going on in each picture. Discuss what Coach Blue writes on the sign at the end of each practice. Ask "How many laps did the whole team swim?" and "How many more laps does the team have to swim?"

• In this story, Coach Blue uses the traditional U.S. method of subtracting. On the right you can see some other ways to subtract two-digit numbers. Encourage the child to experiment with these strategies or develop new ones of his or her own.

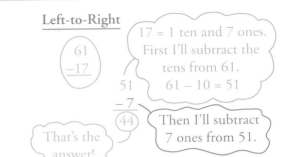

Left-to-Right

$$\begin{array}{r} 61 \\ -17 \\ \hline 51 \\ -7 \\ \hline 44 \end{array}$$

17 = 1 ten and 7 ones. First I'll subtract the tens from 61.
61 − 10 = 51

Then I'll subtract 7 ones from 51.

That's the answer!

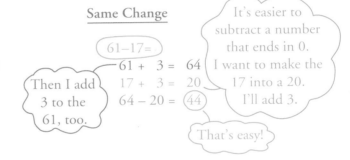

Same Change

61 − 17 =
$$\begin{array}{r} 61 + 3 = 64 \\ 17 + 3 = 20 \\ 64 - 20 = 44 \end{array}$$

It's easier to subtract a number that ends in 0. I want to make the 17 into a 20. I'll add 3.

Then I add 3 to the 61, too.

That's easy!

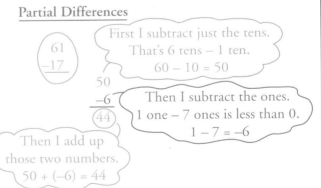

Partial Differences

$$\begin{array}{r} 61 \\ -17 \\ \hline 50 \\ -6 \\ \hline 44 \end{array}$$

First I subtract just the tens. That's 6 tens − 1 ten.
60 − 10 = 50

Then I subtract the ones. 1 one − 7 ones is less than 0.
1 − 7 = −6

Then I add up those two numbers.
50 + (−6) = 44

ollowing are some activities that will help you extend the concepts presented in *Shark Swimathon* into a child's everyday life:

Calculator Game: You will need two players and a calculator. Enter 101 on the calculator. Each player takes turns subtracting any number from 1 to 9. The first player to get zero for an answer is the winner.

Family Trip: When taking a trip in the car, have the child write down the miles on the odometer, and periodically calculate the number of miles traveled during the trip.

Money Be Gone Game: You will need at least two players, 8 dimes for each player, about 50 pennies in the bank, and a set of cards numbered 1 through 15. Each player starts with 8 dimes. Mix up the cards and place them facedown in a pile. Taking turns, each player draws a card and gives the amount shown to the bank. If the player does not have exact change, he or she must exchange a dime for 10 pennies. The first player to get rid of all his or her money wins.

The following stories include some of the same concepts that are presented in *Shark Swimathon*:

- PIGS WILL BE PIGS by Amy Axelrod

- ALEXANDER, WHO USED TO BE RICH LAST SUNDAY by Judith Viorst

- HOW THE SECOND GRADE GOT $8,205.50 TO VISIT THE STATUE OF LIBERTY by Nathan Zimelman